Worn Smooth
between
Devourings

ALSO BY LAUREN CAMP

An Eye in Each Square
Took House
Turquoise Door
One Hundred Hungers
The Dailiness
This Business of Wisdom

Worn Smooth
between
Devourings

poems
Lauren Camp

NYQ Books
The New York Quarterly Foundation, Inc.
Beacon, New York

NYQ Books™ is an imprint of The New York Quarterly Foundation, Inc.

The New York Quarterly Foundation, Inc.
P. O. Box 470
Beacon, NY 12508

www.nyq.org

ISBN 978-1-63045-102-8
Library of Congress Control Number: 2023931418

First Edition

Typeset in Gotham
Layout and cover design by David Camp
Cover art: "Truchas" by Fran Hardy (www.franhardy.com)

CONTENTS

"The path of water is not noticed by water,
but is realized by water."

— *Eihei Dogen*

Worn Smooth
between
Devourings

A PARTIAL LIST OF HERE AND FAR

Out in the sky, no one sleeps. No one, no one.
— Federico García Lorca

Acres of weeds to our view and every minute
needs reminding
that we haven't yet been erased.
A birdcage floats over
a scumbled sky. Just another day
bothered by its own matter and sometimes
this is enough
of a glimpse of what's left.
We're living in a summer
thrown to an oven. Soon even
the woman painting the blue
of her roof has grease on her fingers.
A man stirs a pot and the town
tips sideways. Who are the innocents?
We evaluate our mortgage
to see what we owe
on the trees. Turn flour
to muffins in the middle of a great
battle we're having with disappointment.
A sprung center of spit-back hollers
and harrowing to refigure doubt, then a late supper
handed over in crinkly paper. We settle
to gowning in. One of us admits the thin night.
That's what emptiness has become.
There's a constant gristle of air.
A stone moon parties the dark.
Around us a perfect landscape of ruins.

ACCIDENTAL SINGING

All my tremolos rise past midnight
and I carry them down
the hill to my foggy

little corner where I store
some of my disorder. Standing
on the periphery

of light like this, I am able
to think we go on
from our addresses, absorbed

into what is less like confinement
and more our frail
territories. We're glad

to be reminded to succeed
in our staying. To sing
the inevitable echoes.

Such an album of prayers
I've carried across
the country: from my lonely

New York into Boston
with its chosen darkness. Outside,
right now in the weighted

desert with its endless
edges, I arrange all my
histories, and the moon

is eternal. I wait for it
to become common, and while I wait
I whisper a song

I learned from a woman
in a slow office, 23rd floor—
a woman without features,

just a white head scarf.
She knew grief, but didn't know
she was teaching me

a guarantee for getting through
every next day: some words
in some order, those words

I could keep until I found
my way home. The sound
as steering, always beside me.

GETTING TO WHAT I KNOW

I am desperate to praise the pleached yuccas
on that corner. A red bench and beetles. A labyrinth
moving a thousand arounds. I've never looked
at our village this way: past the economics of consoling
low orbit, the strata embroidered, breathless
crevices. I'm so often simply porched beneath mountains
of light. We've earned a trip to Lisbon or Chennai
but can't make ourselves crosscheck the boxes
to get there, can't quite exit this marbling sun with its languid
pinks and migrations of cranes. Here, everything ghosts
into terrifying lusts and gutting wind; everyone's
rattled. Juniper pollen, absence. We workhorse for hours
and flippant exhaustion. Pull meat from the grill
then scrub tines till our arms ache. Rinse off the cactus
when rain won't give its crystals. But four years ago
we flew to the southern stone of the Americas.
Beneath clouds, we said little but folded cancer's ash
to weather-carved ruins, to a constant that smelled
of elaborate monkeys. The moon was sipping
moist air. We sat with our empty cupped hands. Looked
at salvaged maps to find a next path. Walls, castles,
city buildings, beaches. Crucial rips down the center
showed us direction long as a wing.

PROPERTY

After each lyric drop of rain   falls   bodiless, shuddering
      and rippling   to the shoulder   of this parched earth.

After this,
the deep-throated sage, artemisia and juniper slowly lift,
arranging their scent.

A desert takes what staggers to it.

The storm landed   in unplanted pathways   the spiritless
          withered places, transfixed nearly
       to stone, and now the ground
           is socked in a blanket of flat-patch goathead.

Winds hiss through, unfinished, to say something we don't understand.
What we've planted fails under the branchless sky, and the periphery
of the property is wrapped in fast-formed stickers, a crowded geometry:
precise, spiteful, yellow, without margins.

Each morning we scoop with trowels.

I had never loved a land enough   to want to bend
and whittle out the dangers, to lift them up by centers,

needling the soft pads of my fingers where they gaze upward. Enough

that I would sign my name to each spot I clear
with a drop of blood.

My bucket fills with five-sided thorns sprawling like stars.

            And in the end, nothing left
                  but the dead-dry ground—
            again shredded at the effort of pressing water to it.

DAYS OF THE FIRE OF FIRES

We said—here, too.
We said no power.
We said oh no. Said
many prisms of pity, conclusion. We said
those short and small messages
that mean terrible, horrifying.
Can't believe it we said,
awful, and thinking of you
again and again. We said wow
and OMG and nightmare, and
we said have a plan, said love and worried and hope
you're okay. We said plague and go or don't.
We said no clearing of sky and more
than a week. We said vast
mist, the color of rust. We said as if
refusing what can be undone.
We said forest and town and erasure and species.
Some said we are still okay.
We said sunrise and ash. We said same
here. Or we said never. She said trying
to breathe. The winds rape on
in the unfocused light. We said it is,
it isn't. He said the full coming down,
the coiling around. Unreal. We said ravaged,
so very, I'm so. Sorry and so. Same here
we said, or yes or no. Tragic we said. Look at it now.
We said this morning, said we are settled
much as we can. Said our house
or the neighbors, everything, everyone.
Likely to get worse we said we heard.
The end of the world. How are you?
Pray we said in a whisper—
the valley is heavy.
We drove through. Could not.

LESSONS FROM A TEMPORARY NORTH

We left our house—
the stucco cracks, left cactus crown and spine, arms
out of which flowers came
like savages, left our spoons
and the mounding music that sleeps us
into anviled night. We had eaten dinner: a bowl of rice
with yellow butter, and drove until only solid light
wrote our way, the wind a snap
between junkyard and mountain. We drove
through red wings, crusty forests, past wood
fences, and arrived where houses clasp
to each other like shells. Our language after
this day of transition was becoming
a negotiation, a nectar, a table
of day-old irises, the window
of a bedroom. Outside, a white dog's rough tongue
tasted evening. Ambition: the accumulation
of comfort, leftover cake. Frida painted
the succulent cut flesh
of watermelons on her last canvas. *Sandías.*
Three years later, Diego, god-warned to his edge
and fluid filling his lungs, painted
the same earthly fruit.
There are no casual insights. We dared
to search out the seedy
lean of the sun, looking both ways
to its softening. Within semicircles of earth
we found comfrey, horsetail and plume. We found calm
and lull, another contrition. Fields
with no meanings but wildness. The windows
of the cabin held, inside
and outside, two particular spaces.

BACKWARD TO SLUMGULLION PASS

We were about to reach the sky—to our left a full plummet
        and trees charred to graves, a pulse
of ravens sketching land for dinner. In the bones of this, we saw
        tracts halved long after they were loosed
from family, a ranch intersected
        by bridge. Bridge separating consonants
of old and new. Grazing cows. We didn't need
        much or at all. My man might have considered
happiness and how we have had it
        at times. He had that silk look
of vulnerability. We were letting our silence
        grow and vanish and wake again. How human
our hours. Begin again with only this
        to do. We hiked. Our bodies called
their limitations: foot, hip, more. We followed the veins
        of soil. Landing in truth. Lizards reeled
on stone and slipped through edges. Birds pecked
        a melody that confessed nothing
but peaks. Now was also then: undeniable.
        We kept climbing. Stone, stone and arm
and maps, a circumference of tracks,
        of unbreakable heat. I stopped thinking
about thought. Simply the same red of the cliff
        that had always been. Simply the violent height
of exhaustion. Some kind of sweet.
        Each length of dust, the root of details.

FOR SALE

In the spectacular town where I live, the sun offers
       forty-two medicines every hour and I've
       never been somewhere

as intricately cinnamon or poised. Most people here talk
       of planets and divisions
       of retrograde. We all need excuses. I don't

have even limited knowledge
       of astrology, so I just now Googled
       and learned the sun controls ego and self, and today

my husband quit his job with the angry
       realtor who runs in and out of his downtown
       office and stages his frustration

about bank details as he jiggles
       cold keys. My sweetest left him an invoice and *Sorry, I'm done*
       with this attitude in the perfect cursive

he's used since fourth grade, then got in his truck and went
       to the corner for fuel. Saw a homeless man squatting
       in a small diameter with a gas can

and guitar, a grin. My man filled that little tin tank
       and drove Carl across the street. Carl wants to start
       a circus, my sweetest tells me, and I think we all want

to conjure new ways to loop
       from daily rituals, the muscle of pandemonium
       and anger surrounding our heavy times. Want bold

enchantment, the sugar of thickened color. Across town, right then
       I was at the gym: left, right, obedient
       core, and after, in the locker room, talking easy

to a woman changing into a robin's egg
    blue swimsuit. We savored tales of the waists
    of our gardens, my slow timid

aspens with their tiny gold bells. She said she had extra Theves
    poplars and do I want them. Yes, I want them I
    said, because I need tall

and stable in this life. After all, isn't there always a cannon
    of news? To evade it, I again go online
    learning what seems ridiculous: ranks

of sentences. Declarative, imperative,
    interrogative, exclamatory. The very substance
    of dare, of wound, beginning. No one knows

which bends the most. In my rural village, a neighbor is selling
    tiki torches (10) and bottles
    of fuel; someone else has a picture of multiple

tires on Craigslist. Everyone is selling
    something to someone. Everything is switching
    hands. Did you know there is new violence

in Gaza, people throwing burning tires and rocks? How we tell time
    has a perimeter. Around us, shifting
    performers. An open air prison. Filled with its sentences.

ECHINOPSIS PACHANOI

Tell me from your mouth what can be saved.
We cut apart the cactus with a sharp knife,
alcohol, mid-drift of thorn. The weather's turned
to a mist of heat. Awake, asleep.
The stems want only a bite of water
or rain aslant. Today's work
is dry weeds, a roadrunner slicking past yucca.
Sometimes when I feel betrayed, I need to lean
against the desert, to remember desire
isn't the end. Three-awn, kinnikinnik,
pine-drop seed stalks, the metronome
of sky, steady along.
Am I missing the answer: light
through its rotations?
Before I get old, I'll learn how it sings
nettle, scale, the spoon-breath
of dust-eddy, sip and remnant.
I will have seen a whole translucence.
The cactus insists on blooming. From joints,
eyeholes. Meaning, what splits from this
gives me the fortune to open
in welcome. I must want
what is empty. Map it, mention it, talk
all day. The month folds its tired self over me, clasps
without moving. I kneel
on the ground which keeps on
surviving. Above me wings with their rituals,
a sequence of hours, chance. All it takes
is summer again. Another morning.
What it hasn't been.

WHISTLE DOWN THE WIND

Everything can't be as large
as consent. The man I like to forget whistled. He had been
hawked out of paradise, or he never knew

paradise and his whistle chased me
to distance. The clock at the church kept to its ticking, its bells
with their sticky, poised hope. I could not withstand

reason. That year
I gathered noon every Wednesday and started again
every noon with the sun enthroning

my heart. A maze, a pace. I have no excuse
but the sip and the fork. Let's stop talking
about sinners. Aren't they everywhere? I give credit to the useless

memory. I had to leave the calendar
to find sun familiar again. Now I gaze at this
quarrelsome desert: barren

with discipline. On Thursdays I go through exponents;
on Tuesdays I imagine needing
only the line. Some days my days

are identical. Shame
is expected. Shame is easy. It's easy
to show you what is left of desire: freeway, curb,

quiver, flight. His gift was the dust
in which I breed seeds.
Soon, time became a thin piece

of past and not necessary to keep. I consider
what has happened since: the six
hospital visits, the episodic

arrangement of yucca, our ruined America. And so on,
because it is all a repetition
of loss and immensity. The man I cannot define

never abandoned his hazard. His
circulatory system. And it was this that I vultured
for unboned months. More grasping

and then more farewells. In the garden, my madness
showed up as thickening mint. That wasn't enough so I made myself
to bear others. I don't know why

I started to think about this.
Everything is concluded. But there was such aching
to hasp what I thought I must have. It seems I will never dormant

those days I stole meaning
and arranged it like love. The weather was rubbed
with fragrance, the yolk of the blistering sun

practicing its offerings. It was every degree of
fun or appearance. Those noons I expected to eat and I ate
and starved. The clouds spilled

over the sky, coupling.
What matters? I think through every aperture. He gave me
nothing but the jewel

of darkness. I wear it a lot. The coyotes
are out, compassing
the ground. Do you hear them? Whistling about their beautiful lives.

DEAR DESIRE

I don't know how I built this
need. Where to hold it.
I cannot explain why the wind inhabits
my body, and I would like
to see more gold leaf. I would like everyone
to see it. Last night we watched another movie
in which love is a threshold and the city
around it, deep snow.
What I mean is, I'm lucky.
I post magnets on my fridge. Small print
stuck on a syntonic grocery list:
tin foil, cauliflower, salsa, a whole journey
of feeding my family. Those clutching
reminders with devotional practice.
Mornings my husband grinds the coffee
with pro-grade earmuffs on.
He wears a blue shirt again.
Another day dressing in distance.
He watches me linger beside the sayings.
Something to use between the world and my will.
What do we know of contentment?
My husband is exhausted by the unavailable
trouble I crave. My magnets tell me to expect nothing,
to never stop wanting. They repeat every day.
After this, he sorts the laundry:
darks, lights, blues. These are the rooms of our house.

THE NET SIGNAL REACHES THE BODY IN A FACTOR OF 2

I count 32 robins in the dead tree at the curve
of our road. He counts days since rain. Every moment we assess
the end of things, the knives we need
to tender the meat for dinner.

Before I leave, I put my face to the mirror, realize the long bones
of heaviness and the bitter things I've forgotten.

On the drive to work, at a small corner off the Interstate
in an echo of sky, thirsty,
wheeling along on small excursions,
there are racks of army fatigues.
Many people shop. Sunglasses.

I keep following directions
to the shadow of cottonwoods.
All the hope in this gold will erase.

I don't lose sleep because it is winter and already
full of its letdown. The weather keeps coasting through,
rain plucking the ground, or the little shapes
of dry air edged with olive trees.

I hold nights in the little spot at my hip.
I hold days by the wrists, wrest hours for loving what doesn't exist.

Next time I drive to town, I see the top of the mountain
is missing, dusted with gray
like a grief. Pinholes of light in a flat morning sky.

Is it still a mountain if it is
locked
behind itself?

I see my world in verbs, and say it
mistaken, hurried with ongoing action. Say it as origin

and corners. My love returns
from a cardiology appointment where he's taken his compass

of anger and each consequence
we can't determine. They've heard his heart
beat a pace that is four times the normal.

And so now I put my ear to his chest like a button:
I am sewn there, listening,
holding on.

WHITE CHALET BIRD HOUSE

Where do you house your worry? At 4 am, I'm awake
to the surly news kicked up in the dark. I'm alert to the wind
that keeps situating. The season
I've grown to detest now goes wiry all year and every
hour I am reduced to its lacing.
A falcon stands guard, toothing the air
from its perch on a dead tree to the east.
I have so much work to do
on the irrational brutal future
of the world in these murky hours, work that never
is needed. Or finished. Instead, I begin
studying theorems, blink through the language
of lemmas and identities, and bother
the buttons that break me from such
propositions. Maybe math is not
where I need to be. I used to be grateful
for proof, but now want something less narrow
to settle in. Do you worry? I've been going
to friends' houses and looking devoted
for radiance, sitting at their tables, in their barns, running
my knees against their rivers and petting
their turkeys, checking where they are safe,
what is lucky, and what might be broken
to sorrow. On the Web, I find a site
with multipart birdhouses. Rooms to move into.
Can I say housing keeps me up?
I read of the stabbing in Portland as people were traveling
home. Then, the busful of Coptic Christians in Egypt
en route to the lodging of their hearts. I cannot sleep
through any more slaying. The birdhouses are all white
with shingled pine roof and windows
and other little advantages: a porthole, a door.

A house is for safety, for nesting, for the songs of comfort.
I read and read through the stars and the balcony
of light outside turns beautiful and the wind
is now easy as a spoon and not writhing, and the nerd clock
on the wall plans its way forward with equations
I'll never understand, with time
that will always be indefinite, that might even,
when I'm not looking, escape.

MORE REAL THAN RIGHT IN FRONT

I put on a shirt inside out.
I drank the coffee.
I put on the shirt again.
I checked the drawers, kept all the rotten
plots and exhausted timetables.
Motion or salt, paper or fist, pocket or hurricane.
I had abandoned my teapot, the broken chair, the mountains
for a panoramic circuit of slate clouds and highway veins.
I would rather the third or fourth day
sun had clasped to my legs. Since there was no calm,
each of us battled for tickets, vouchers, toothpaste.
Fluorescent hallways with their pasted roses,
the iced rooms, many Bibles and desolate lace curtains.
I would rather birds
but instead the airport and shrines to luggage,
then a quick cheeseburger in an overbright corner.
Look, a storm is forming. Some frequency of windows.
We are never leaving.
Another long hall, another
limited use for the body. Concertinaed gates.
Now the sky-signature
of doubling. A window again flowing with rain.
A woman washing her child's eyes in the bathroom sink.
Others walk by rephrasing most common words.
Around me only the universe,
a wing toward dust. Someone turning.

PRESENCE

From a plane—hovering,
the earth fragments in a grid of brown and green.

Every time I look out,
all absence and new patterns, repeated.

I have packed in *shoulds* and come farther to wake
in desolation. I don't know this yet.

What I can't see is duration,
the perpetual upheaval and sinking.

Water chases itself, crying its name.

MY PALMS LEAVE BLURRED MARKS

I tell no one how I compose my days—the calendar of sand
and the math of its accumulation, arc and crashing. One day

pouring into me. Another made of wingspan, made of rescue.
In knotted sweaters, I move over rocks toward the rats,

past the ochre skyline endlessly cradled by the water. I learn
what demolishes and drifts, what loiters,

the water taut by shifting. At low tide, seaweed rots and knits
to benches. By the iron-skinned beach, a bright blue boat

is waiting. I fold firm stones into the belly
of my fleece jacket, carry the weight of what I find.

I don't raise my eyes until the terns cry,
until the litter is an effigy I've kept from drowning.

UNIVERSE

Either the sound is pausing
    or the sound is all
breath, many breaths, the gristle
    and cleft whispering
rough in a rush

    past homes knuckled
to one single plank
    as the ocean snares
deep folds. As each wave
    combs the back of itself

beneath sky. At high tide:
    a great span of white
pelts the shutters.
    A stark sun continues
retreating in small gray

    rosettes, picking out piers
as the waves unmake and exhale,
    twirl again, ease away
over a universe where ocean
    flounces the town.

WINTER OF TUMULT AND ARTIFACT

Again I navigate to the throat of the ocean for audible
guidance and to acquire primitive details from the fraught
plummet and slender horizon: a chain link, a stained
Bible. I make peace with the innocent wielding. At the edge—
a teaspoon, a sandal. A shearwater circles without anger
and lightens, its wide wings, imperial, spread over
the repetition of distance. The wind leaves its motion
in filigreed sand, while out at the midpoint, an elliptical
brooding. The water is full of itself and going nowhere
every twelve hours, and I might be the only one
to believe it or to be frightened. It commingles its salt
with lost objects and spits out a metal button, the hook
of an earring. The holy knife of water slices
further. Not that this could be simplified to the in-
out of action, the efficient carving of portions. Not
that anything is all farewell and return. Or all hunger.
Every time I capture a hollow bleached bone, a yawning
plastic bag, another token, who says there is focus?
The grim light splashes up on me daily as I find
the remnants of strangers—a small tin, a key
still almost magenta. So it is pure reason to tend the endless
wealth given by a thousand angles of light as water
dances the riptide. Every day an epiphany of loosening
tedium. Out of the ocean's sleek windows,
every vulnerable object seems to surrender. What I clutch
is only a resemblance of yesterday's losses claimed
by the ocean's euphoria, now docked in the slack at the shore.
The cold is still luminous and preening. In the end,
there is no end if you stand long enough beside it.

THE FINEST LIGHT

A stone wall
against a bruised sky,
a red mailbox

and clapboard saturated
with morning rain,

sun-haunted lilies
beside a worn bench,

a day of wet wool,
algaed graves on a tired acre,
the strange perfume
of trees,

towels on a clothesline,
an old woman
on a bike with a basket

and the same rider
returning
when the sun is heavy.

FACE, CIRCLE, ABUNDANT

My friend Virginia said I would miss the dry air
in those months between places.
At home, my apple tree had toothed to the earth,

limbs already browning. At home, a red-faced sky. Everything
was anchored, fluted by light and my fingerprints.

In this cottage, I found my smell between broken utensils
and the slap of glue in the basement.
It wasn't that I wanted to be alone,

just that time had been given amid plank
and dunes. Just that I needed the underbrush, the container,

blossoms of observation. Needed to
aspire to the wisdom of waiting.
I kept an old jacket, a flowerpot, and a field bag

with a toothless zipper. Wrote words
on pieces of cardboard:

> *monochrome*
> *face*
> *circle*
> *abundant*

I studied the shore, its hem
at the sand. Perhaps I was eluding myself,
a woman wrapped in three coats beside multiple rough ridges

of ocean. When I wasn't apprehending the deep sink
and succession of water, I trusted daylight

until that, too, dimmed. A quiet emerged
through my palms. It was sewn
or written down, which made it less singular.

Down the road, the ocean yanked
from its cashmere center to write rust on old objects.

Waves spun, slobbered, disappeared. All that weight weeping
onto shore. My friend Virginia is usually right,
especially since she moved away from the desert,

but I was making a language each time
the ocean shouted and bulged.

And all the mornings and twilight, I looked through
its doorway and listened
to perfect circles bend without breaking.

ELSEWHERE

There's always a reason for elsewhere, a chill, a chance to leave creases. At the narrow end I lived with my dust in its traveled colors in the bones and rooms of a clapboard house stained a sad unpredictable red. The house had been floated over the harbor, and sat, solemn, one street from the mouth that never ceases. In the mornings, as I recall, the flocked listless light cast a hull on the back wall, and by night maple leaves hovered like fingers in darkness. Each road agreed with the next, taking me forward. The hours in each cycle rose and receded. I was bystander, hidden. Water sprawled its salt-broth against dike and docks. The ocean undressed to an exquisite blue gown. Day after day: empire, departure, routine. The hurling Atlantic throwing its gesture on sand. Autumn layered and pressed against me. Even now, years later and back within the desert, the undulation has not diminished.

RUNAWAY

Olive ground, quarried
        ground, can I remember anything
        green? If there is no far

        away, the wind must be our revelation. My love stands
        beside me. We are connected
                in silence in the sway of a bud

        of thought. I expect his perfect words
of distress: trees to ash, fence line, rabbits,
        rats. The politicians do their damage.

                To cope, I plunk sand into the pickup.
        Drones, boots. We are already home. I glance across

the barren, build my small
                mounds and blisters. Step away

        from the news. I must choose
what matters. Some people say their old sayings
to pace the ruin. Some wolf and threat

        against friends. What I need and see
is the cactus pads, endearingly succulent. Also true is the sky stops
                every violence. Despite everything

        this morning, I ordered the cupboards
        to jars of rice, noodles, beans. Ready valleys and hillocks
                for when we'll hunger again. Now the sun with its egotistical

joy atop clouds.   All day I will vigilant
        the insects, buzzing in their endless

campaign to track down an oat, a drop
of honey, a little shear of flesh.
They need only the simplest. Not a swarm, not

terror: see how fallible
I've become? Coveting the negligible

world and exclaiming the ache with my eye
to the ground. Cars along the freeway
a mile off make a resolute noise. This is the size of my everyday

concerns. The desert fills with its drastic
reduction. Scrub grows. A succor for some. I bend down and pet it
until I see Adam's face.
Eve's. How it started to ruin.

TO REACH BACKWARD

Back in Arabic, my uncles pitched time to its simple anchor
of wounds. I couldn't mine their shifts, the crisp industry
of anger, so I tunneled through food, which was plentiful
on white plates that ached from the buttery weight. We were
in the basket of another Friday in the suburbs. My grandmother
in full apron and arthritic fingers. She heard no compliments
unless you count how well I ate lamb's tongue, which she'd stewed
for an hour. The food came as history told her to make it; the food
came with its vermilion river of beets. The food, spiced and rich.
With its lost location, that food. How long were we still hungry?
Each bite, loud noises. Remember how the room hummed, the voices
swaying the ceiling. We all understood this was now, our palms
moving fast to claim stews, which were bursting out steam.

HISTORY OF SALT, HISTORY OF SPIRAL

It is said the man built a spiral in salt and in six days.
He wore hip-waders. Waiting.
Ordered rocks. Made it to hold the soul
of the water. A backhoe scooped and anchored.
Beneath the crown of sea, hoe on hoe
had piled to a pattern of metal.
Archimedes made a spiral to place a curve.
A curve can run from a fixed point to a length.
The artist rebuilt his spiral. Wider,
built it to circle backwards, from empty
to the ambit of cosmos. To place
what wasn't there. The backhoe carried basalt,
placed it in salt. The salt was pink. The flats, green.
The salt was white, like nothing. For a while, rocks
were buried in bubbles of salt. The man in black waders.
The mathematician imagined a spiral that threw
constant speed, constant velocity.
The artist made a spiral to match the horizon.
The lake is shallow and without outlet. Ice age lake
watched, unwatched. Within the spiral, you cannot find it.
It is said the lake is nearly one-third salt.
The salt makes a crust. The rock is black. The spiral
is always a spiral. You are there and not
at the same time. And you are carrying
a question. Maybe all your questions. To walk it
is to accompany dreams. You walk to a sound
that is neither a hum nor a squeal, but time shifting
its angles. Sand seeding. Later, the colors
will be earth, pink, green. It is said that absence
will come forward. You will need to see
from above. You will not see.

THE WORLD I CAN'T REMEMBER IS NOW

My friend has disappeared though he lives
down the road between sweetgrass, updraft and evergreen
in a made house I've never seen.
We're living in a puny monsoon season drenched
with waiting and in starlight totems of sky.
My friend has hunkered down with his four
little dogs' ceaseless fur.
He tells me by text he discovered "will have left"
is future, present and past all at once.
Which means I will never be through with the eyeholes
of my top floor West Coast apartment
where the fog whipped limpid and filled with azaleas.
In such atmosphere my shadow learned
to burrow in disorder. I've also left and am likewise
still bent to the fat filthy resin
and liturgy of Manhattan, and the runes and disbound fields
of Montana, those weeks without heroes,
those wrecked weeks I gladly
went no place but the pond, and yes,
I also lean in now and then to the orbit and habit
of Boston where I hung in hounding heat, broken up, shivering.
We are suspended in places
entire and different and home.
What I left was the restless growing, the slippery
wages, each white tread of evening. Left witches,
sieges, drafts and domes. What I left
was my tender body that wanted to punish.
Look how we adjust to such up and leaving.
Though I remember porcelain, soil, less flesh,
I have to reach backward to do so.
I trust my friend who is absent, my friend
who gives me his eye every two years:
making photos of my face like a meadow, tilling my body

for abundant light. I tell him last summer I climbed
the surface of a small mottling in the Organ Mountains but left
all the bones in their places, pressured to earth.
My friend down the road texts many times. Today
is the same as before, and we quick
question nothing and that's what happens.
I find a book on a shelf and now the rain has begun
and is tilted with promise
but this is the desert which means
by the time I say I felt a drop, it will have left.

A LONG STOP

Many neighbors live here at once, tawny
homes, floorboards, sturdy knotweed, rats in the pulp
after sunset pulling the wrung
laces. Indifferent erasure. What is wrong is left
to source. Side yard, shaggy hillside. Hatch,
months, wounds worn out. A train takes
its thumps—a steel cape
to the landscape. A girl walks a ridge and might be refusing
the research of copper fluffs on the scrub. Or sooner.
Or later. Might be practicing her wagers
and desires or rickety capitals. Advancing at least.
At night I read the news aloud in bare feet
and remember eating enchiladas
in a cemetery within calculated textures and green
initials, no roam. All the dead
lying bare in a flat catch, one pear tree
loose, leaning. It was May
in El Rito with its rust and colts
and a bridge dented over caked water. We got home,
backside of the freeway, through all that
origin. In shadows
and stirring. Now all nights
hum with the abundance
of statistics, conditions
and paupers. I'm stubborn
to find a nothing and scroll until the end.
Whatever he could reply, he rarely does.
I write a love letter to the owl who slipped
from the giant cottonwood
by the freeway when we were walking toward sunset.
Gallant recusant. We were walking away
from home to walk back

soon. He was breathing harder, older.
And now no names are drawn
from aftermath. We must level
to recent and not long ago in each marketplace.
We've been together dividing each day
to ritual: dark coffee, the fluid motion
of birds. I figure to know who pits our truths
but every color during winter is borrow and powder,
is leather and leavings,
the vowels distant, the whirl
of clocks lost to some orbit.
So many half serifs of prayer.
The hand of questions has climbed into our bed.
Looking back, I realize we always have light
even contained in our medley
of poverties. We've never adorned with it.
Who is taking roll? Who is taking photos?
This year we planted a tree and another tree
and took out three roses and the blood
of our worry, waiting to see.
We may be safe.
I do most of my beseeching at dawn. I recite
all we have not been.
Partway, the sky starts to trembling
light: thin and pink.

GESTURE AND OMEN

Yet you have gone
to Nunnelly while the country
is deep in its minuses
and the nurses are dying.
Praise the day you tired
of crying every reply
and mesmerized feeling.
I reached out to your children, the one
in his humming
apartment who has
gleaned a raise. His hair
combed. He will take you
to Ireland when we are again
free to go. I found
your daughter
slipping around
streets with her palms
out for money or pills and a beach
nearby hollering
its rings of inattentive
water. She says she is scared.
To listen is to hear
what her body wants. A place
to shell while you
are in Nunnelly, which is a place I see
on the map mostly
cloudy today and a road
called Elliott. Are you getting
better? I want this pause
you are having
because here
we are in the midst

of the worst. Outside is
no cars
and no mothers.
A list of sad doorsteps.

ACROSS THE VICTORIOUS
SCRUB BRUSH, CROW SPIRALS

shake the sky while I walk back for hours
in a particular emptiness, air hoarse, fitting

to its broad-shouldered detours. The land magnets
to rises and I move toward the high language of tree in a starved

arroyo, a big cottonwood that I once saw twitch
with an owl, voluptuous, who dawdled until I was directly under

pillar and leaf, then repeated its wings and was off
with a vision of other fault planes,

pulses. When I come out here, I give up
all the fictions of the nation. I can taste evidence of silence

and hold my thought. Just me and an unruffled
verdigris loneliness, which is maybe the comfort I've best lived with

as normal. Where the desert bakes its stones
to old oaths, it is easy to count all the conclusions

as seduction. Right-sized, half-forward. Coming here
means I can see if the tree is blue with raking shadow, star limbs

and leaf-taking today. Means landing. Dreaming.
I don't know if this is proof

of mortality. But I'm whistling. The end is always near but every future
has been forgotten. I was a happy girl. I remember so little.

A GRAY CORNER FULL OF LIGHT

People around me pray
for an ease to an edge or the hopeful
tiny birds in a nest or any other existence. Everywhere, hearts
building heart-like behaviors: stomping
and sadness. My love says I'm over-invested
in the parade of habits, but I look
at folks' motions and dogs and the way they can't
for weeks sometimes
get out of bed. Life wounds and stars.
Terminal glinting. I loved being a kid, living inside
my name, studying for stupid exams. I loved
being loud and dependent. Someone fed me or stopped me
overgrazing; someone told me I must
go to temple again
and I went and I went. Uttered
the scripture of inquiry. I sat beside humans
in hats to sing kinetic adoration
of an almighty in a language filled with the choired
muscle of moaning. I was awkward. I was young. God let me
be what I'd memorized. The thrust and breath of it.
On the radio last night, Jim Carrey
said that Buddha said we are not separate
things, which he believed can help cope with the lurch
and gust of the world, and then he laughed
about making 10 million dollars
on a film. Muddled silence. I keep bookmarking pages to read
about Rabbinical teaching and Gnostics. It all comes back
to strategies. As a girl, I practiced my prayer
on Saturdays, Sundays, Tuesday afternoons.
I didn't know we needed salvation, but God
bounced through every sequence. For sermons
I wore dresses. Faux velvet. I stole gumpacks
after each ethics lesson from the sweet

shop. Made of a hunger. Inside me, what mattered. Now I pray
for the icecaps, the bruise and wings
of today's children who will grow
into this regrettable culture we're busy
to empty. These days I keep the eternal
within my mouth. Mumbled
questions, split lip. When I'm scared
on streets with missing pigeons, I pull up those vowels
and start the work of worship.
The air whole. Something seen,
worth praising, something about to vanish.

TELL TIME

Through that long stretch of knuckled nerves as every inch
of our property dried out, we saw no one
but the wind quivered, tearing circles, and my love
cut boards and built a box
for the worms to lurk. It wasn't over. We were
alive. We planted lavender against the house. Our prayers split
to outbursts and technical questions. Every night news
narrators muttered. I said to David I want those stars
in our trees, the oxidized Mexican tin
wrestling the tufted junipers. From the roadside
we saw the pulped pink moon as a whole fact.
If you covered one eye, our house had more lizards
than windows. One of the few days I went out, I bought
raspberries that mashed. At home, I took pink
handfuls and slipped off to read a big story; heard no one
but pages. We were safe and still David hardly
slept. We were safe, but days after we unstrung
the clothesline, we hung new rope. We fixed
the cabinet doors from slamming.
My molar continued throbbing. No one would pull it.
I lit a hundred candles for a likelihood. The passionate
swirl of the ravens. At every coyote fence, I peered in
and every ditch and so forth and I dawdled and was
absolutely unknown so I stumbled back home. And stayed.

I'M ALWAYS NOW STUDYING THE URGENCY

We moved here to live, and we didn't know
the land. We moved here
early and to this day.

We heard the fluted salt of ridges whose job it is to lift color, to chorus
inward our attention. What I couldn't recognize kept flying. Open country.

If noises, I never figured I'd learn to hear them.
At night, other others.
Sheer wonders, roaming.
For years, nothing happened many times.

For years, this was the source of miniature
pleasures. We traveled about, to lakes
and signs and a spun smell of meadows
recalling our senses. For years, fever passed
through without taking.

We moved here

and faced the arroyo and the arroyo continued
to pull its occasional water.

The arroyo opened to a middle and moiled
running. Coyotes trained on this land, spines
curved under gray robes.

We doted on their throated diphthongs
in the plain fade of light, cherished even what
they meant, a coming erasure.

We moved to the edge of vibration
and all the minuses and wings and ancestral gravesites.

What did we need but sumac in luster, halved history and single-celled
layers . . . . what but finches, the leaking desert of light.

The same many times. Rio to rio

. . . . what but aspens greening to nimble.

How lonely was is.
Twenty-nine days since meaningful precipitation.
Hours dust in our lungs, in our mouths.

Keeping our paces and to the side of us, migrations.

A future less wolfheart, snaketail, pinenut,
thunderstorm, snowflutter
and other ordinary devotions—
we couldn't predict this
terrain of ruined tree crown,
flammable. We moved here

to the slow mountains.

Moved here on faith, which is to say what we did, we did
by chamisa, althea, fallugia, blue grama and dropseed
before we even knew these things.

We knew only the previous moment.
Knew to begin in mixed cedar.

While we saw a marrow nearer than farther, the sun bent to fire its own house.

Land braided drought zones
and scrims of smoke in the summers.
The arroyo caught in its unruly shell,
its dry network of tree multiples.
We focused between closely bunched weeds. Endless.

Fire eyes the forests, audacious, dines on some houses.
The canyon barely remembers its water.
To keep from dissolving, I take photos of smoke.
I take photos of slopes. Take photos of clotting.
This is the fifth root of grief.

To prove we won't always be blanched in treeless gaping distance.

In one static-charged valley cast west by raven,
where old burls unfinish, we coiled
past shallow erosion, bare knolls and resin.

We followed directions bruised by circuitous
thrash. Mirror of cloudbank.

Wind came clumsy, and let's be real,
we trifled. Upspun, we didn't see failure
till winter. Didn't see winter
till the white stopped falling.

How lonely is was . . . . . .

thirty-eight days since meaningful precipitation.

We cut the dead stuff. Choke off next branches, lost leaves.
Some people are pushed out to nowhere.

I remember rosehips.

Pick me a scent. Pick me an apology.
We moved here to live in the early morning and later.

To wake to opuntia, a flat faint green.

Four stalled storms this week. The bird kingdom still whiffles dusk in front of us.
Whiptails with their intimate shadows. Aspens with their keening.

I sit by a wall, want for wet, wait for penance.
What you claim as cold,
I claim as tending.

We moved here to a land wrought to hurts.

Inside the warped forest,
I took one picture of worship,
one picture of the brink. One picture to picture it
untucked in long spasm. Shiny vehemence.
This wasn't a pilgrimage,
but a way to remember conductive pulses.
Red scars and recurrence.

Among the ways I can look, I find a river's eyelash edge.
Watch reason lace in small motion.

With fly and lure, chumming and throwbacks
men vault for bass and bluegills, trout and pike and perch
and the river grooms its symmetries.
The river is a yearning, a small seep, a lowest door.

                                        Among the ways I can look, I look at harm.

Coyotes wedge each purple night. Among the ways a blaze smolders.
What happens is a particular peril, heat's speed
throwing time into vectors and aspect and bucket,
to burning period and dispatch.

Heat doesn't request but seizes fences, and we learn

                                        glint, dawn, wrench, echo.

    Still drunk yellow dandelions reach dirty sides of the highway wanting nothing

but public reaction and we do not say *now please I you* and do not
but breathe the stain from which we locate the mountains' ashen edges
and ambition.

The land devours itself. Uphill, upspread.
Lightning, human, arson, campfire. From the inside of window, I go to a map
and choose cinders and incision. The map is all pixels. The map is the size
of the palm of my hand. The map flickers thirsty flares:

Medio Fire	transpiration and flame length, knock down,
Little Bear	running. We moved here. Snag length,
Pacheco	leftover slash. Handline, burnout, plume, sear.
Whitewater-Baldy	Fire and fire and fire cluster and finish.
Dark Canyon	The palette is blackened, a wreck—elusive,
Dry Fire	exposed. Without water in the sky,
Las Conchas	there is fire and fire . . . . dead fuel, lulled wings
Boiler	and we're side road, turned road, aperture,
Vics Peak	24-hour fracture. Colorless wind ready
Cub Fire	to work another violence,
Farm Camp	to fetch and bruise another.

Rough cuts. We moved here to land.
Moved here to live and here
the earth buckles. We are home and why
we can't find our way home. We lived here
late and to this day.

LIVING ROOM

I have watched thirty movies of sheep and fields
spreading slow to the future, and of course I've stayed in my house.
Shredded to the country's mortal glower,
I selected more movies to learn every road and bend
of history and testimonies
of hardship. With the man I love, I watch half
every night after a Manhattan which I never had
until after I'd left Manhattan. The rye burning slightly
my throat. I am as holied
with movies about structures
around the world as with wind bedding the forests.
Do you remember how March
careened? April got smaller. We were already watching movies
then and watched movies with each death that tore
our state crooked with statistics and governors'
warnings and cases and cases. We also watched
movies with squadrons and despots. Blood in all of this.
And trucks and store signs. Every night ill prepared, eating
and not belonging. To abandon the pursuing present, we watch
movies of Malcolm X and movies of walls
in Israel and the sound of economies. All through them
I hold the cat in my arms. I knead her fur and courage. She is dying
and I've known this. Or she is not dying
if I tighten my hold. Each movie comes to an end.
In the last few days, the cat has developed a smell
that I suppose is her organs. My father bubbled blue bile
the final days of his life. After each half a movie, I proceed
to read the plundering news aloud with this cat in my arms.
Beside the living room window the shape of future hours.
We talk of doors. Of obedient problems. Smudges
of work. The movies are ritual and concoction.
The movies are for unbinding the jumble of storms
I've been feeling. And because we only watch half,
they are a fragment or a resistance.

When the cat dies we still watch movies. We drag through half
of another one—about women or gardens or vibrations of time.
On the table, a greeting card apologizes her warm eyes.
Roses the vet sent already unclutching. The fate
of the characters in each movie matters.
I watch for witnesses and tangles,
shorelines and feasts. So many movies.
I think now of edits and widening
spaces and my childhood
where the air was sticky
and hydrant-splashed or porched or frozen
and whatever it did
people threw basketballs
and my family watched as the car went along,
loaded with us. Were we a story?
We drove in the foreground.
Our brown house spoke a language
we all understood. The nameless trees, and on what side
we should sleep. Tonight, when I climb into bed
under the wool blanket with its wrinkles
and my love is falling shut to sleep, I whisper
what I think of the latest movie's fringes,
its motions, the camera angles.

WALKING

A guy with a cane in a hoodie looks down
on the freeway, fettles the shape
of his backpack with his free hand, wipes down
his eyes. He must continue

writing his tired body on distance.
In this hour, nothing but travel,
this evening, nothing but moon. From the moon,
the red of his neck. It is nearly Easter;

people walk, heads down
in the wall of long wind toward the old Santuario,
toward the paradox
of walking without crutches or pain.

Toward our uncontained clouds, they push
through the eye of all doubt—
and if they mumble, it is the edifice
of language: how to withstand.

An intention is set, creased with the patterns
of desert. That I see a guy at the overpass
outside my village means winter's small sphere
is mending, means every green impulse,

every desperate step makes sense.
Pilgrims don't wander into panorama,
but seek a place to be healed.
Wind laps at a road gone loose.

There is something lonely in the near and far sky.
Does he see it?
So many travelers crippled with grief
press to the shoulders of freeway, putting faith

in the picture of particular flesh
and particular gesture: intonations of bone.
Take the blood of despair
from their parched lips. Take their exhaustion,

the density of sorrow, the lack, the imperfect
body with its bristle and promise. Take the sun-
kilned waiting, each bandage and handicap.
Give them luck from the shadows. Some reparation.

STAY INTO

Not the absence of sky but the sudden
work of life: lance or ash, shovel,
notch or wince I begin
inside myself to sing a prayer

for nourishment a feat
of the spirits I was taught one summer
at a wooden table a knot
of bread the euphony

of childhood I memorized
those unwrapped syllables rolling
immersion over years the words
have gone
I take to rhapsodic humming it was

never bounty
but simplicity
a time spun to habit ordinary
inclination a sort of fantasy
god I adored that childhood

those colors I made from my breathing
even the dimly, the stubborn farthest pitches
all this time I've kept
on the lip that ancient song map the divine
worn smooth between devourings

EXERCISE IN HEART

We've had to stop trying and we've had to go on.
The path seemed only soft stems, then a landscape—
spacious, partly vandalized. Birds paddled in
with their voluptuous chanting, slashing
our impudent truths. We were moving along
a ridge. Tender, the sun began to embroider
the meager shrubs which were not flaming with petals.
They seemed to have fallen asleep, nodded off
in the fields but hadn't lain down. A chubby trail
propelled us through the final damp thought
of autumn. Into a quest without speaking
ugly world tensions. No need, after all;
the path sauntered with its geographies
and stony slopes. At one hurried curve,
horses swelled by, faces lined to the prow.
Clop and clop, their purpose; we slid
to the side. The sky in an hour would be
descending in a watermelon lace, quickly wrapping
to a matter of recess, each mountain devouring
the next. Then we'd squint through binoculars at planets.
A quick tilt would show us a halo, a future, a silvery pulse.
Imagine that: whatever we see we can call evidence.

ACKNOWLEDGMENTS

I would like to thank these journals (and their editors) for publishing the poems in this book, some in earlier forms:

Bear Review — "Getting to What I Know"

Beloit Poetry Journal — "Property" and "I'm Always Now Studying the Urgency"

Blackbird — "Echinopsis pachanoi"

Cherry Tree — "Whistle Down the Wind"

Cold Mountain Review — "For Sale"

EcoTheo Review — "Gesture and Omen"

El Palacio — "Across the victorious scrub brush, crow spirals"

Five South — "The World I Can't Remember Is Now"

Flock — "More Real Than Right in Front"

Kenyon Review — "A Partial List of Here and Far"

Lake Effect — "Lessons from a Temporary North"

Lumen — "Elsewhere"

Malpaís Review — "Universe"

New England Review — "Winter of Tumult and Artifact"

Nimrod International Journal — "Living Room"

Paperbark — "History of Salt, History of Spiral" and "Tell Time"

Phi Kappa Phi Forum — "Backward to Slumgullion Pass"

Radar — "Accidental Singing"

Relief — "A Gray Corner Full of Light"

RockPaperPoem — "Presence"

Stirring — "The Net Signal Reaches the Body in a Factor of 2"

Tampa Review — "My Palms Leave Blurred Marks"

Terrain.org — "Days of the Fire of Fires"

Thalia — "Runaway"

The Broadkill Review — "Dear Desire"

The High Window — "The Finest Light"

The Hopper — "Face, Circle, Abundant"
The Los Angeles Review — "Stay Into"
The Quotable — "Walking"
The Texas Review — "White Chalet Bird House"

"I'm Always Now Studying the Urgency" was named a Finalist for the Adrienne Rich Award for Poetry.

"A Gray Corner Full of Light" was selected "Editor's Choice" in the fall 2022 issue of *Relief Journal*. The poem was initially written for B.A. Van Sise's "The Infinite Present."

"Accidental Singing" is included in the Telepoem Booth at New Mexico Highlands University.

"Backward to Slumgullion Pass" was reprinted in *Deep Wild Journal*.

"Property" was reprinted in *Plant-Human Quarterly*.

"Gesture and Omen" was reprinted in Creative Santa Fe's "Moment Booklet: Connection."

"History of Salt, History of Spiral" is inspired by Robert Smithson's seminal earthwork "Spiral Jetty," located on the Rozel Point peninsula in the Great Salt Lake, Utah. It was well worth the pilgrimage there.

Several of the poems were begun at a Gaea Foundation residency in 2004. It took a long time to settle those poems, but the experience of being in Provincetown, Massachusetts for those months has stayed with me.

My gratitude to NYQ Books, especially Raymond Hammond.

This book is for David, whose belief in me surpasses anything reasonable. And it holds within it the memory of Ella, who stayed in my arms as long as she could.

Printed in the USA
CPSIA information can be obtained
at www.ICGtesting.com
LVHW062139211123
764252LV00049B/223

9 781630 451028